# Growing Herbs and Plants for Dyeing

Betty E.M. Jacobs

Illustrations by Kathleen Gough

SBN 910458-12-X

Library of Congress Card Number 77-77943

**Select Books**
**5969 Wilbur Avenue**
**Tarzana, California 91356**

# Preface

My first book, *Growing Herbs for the Kitchen,* seems to have filled a gap in information available about growing culinary Herbs. This has encouraged me to fill yet another gap. Though there are many good books to tell you how to dye with plants, I have found none to tell you in detail how to grow the plants.

So once more I have concentrated on growing needs. I feel that this book would not be complete without some general instructions for dyeing, though these are brief, and only meant to give you a general idea as to methods to employ. My experience of dyeing has been only with wool, so my general instructions are for wool dyeing only. However, in the text (under each plant) I have indicated if the dye obtained is suitable for other fibers as well.

There are many very experienced people in the field of natural dyeing, who have written at great length about their work. I do advise you most strongly, after you have planted your garden of herbs and plants for dyeing, to obtain as many of the books listed in Appendix "B" as you can. While your plants grow, learn all you can about using them.

I have chosen only herbs and plants which yield colors fast to washing and light. It is more satisfactory to use these until you become more experienced, and are skillful enough to experiment a little with others less fast.

My husband and I grew all of these herbs and plants when we had a Herb Farm on Vancouver Island. The climate there was temperate and damp, quite unlike anywhere else in Canada. The winters were not severe, for even Rosemary would winter successfully out of doors.

So, let me wish you good gardening, a good harvest, and finally, successful dyeing.

ALSO BY THE SAME AUTHOR:

*Growing Herbs for the Kitchen* published by Gray's Publishing Ltd., Sidney, B.C., Canada

*Profitable Herb Growing at Home* published by Garden Way Publishing, Charlotte, Vermont 05445

# Contents



# PART 1 HOW TO GROW 23 HERBS AND 7 PLANTS FOR DYEING

## Common Name Index

## Botanical Name Index

# PART 2 HOW TO USE THE HERBS AND PLANTS FOR DYEING

This book is dedicated with gratitude and affection to my Father-in-Law, who raised, spun, dyed and wove his own wool.

# First, You Ought to Read This Introduction!

The information you need to grow and use the thirty plants for dyeing, is listed under the following headings:

**Botanical Name** is usually in two parts, e.g. "Anthemis tinctoria" — "Anthemis", the genus, and "tinctoria", the species.

**Botanical Family** or natural order (N.O.). You will notice a likeness in the appearance of plants of the same botanical family. For example, the Chamomiles, Tagetes (Marigolds), and Golden Rod, all belong to the "Compositae" family.

**Other Names** The same plant may have several common names in one language; sometimes two or more different plants have the same common name. For example, Dyer's Broom may refer to Andropogon virginicus, also called Broomsedge. It may also refer to Genista tinctoria, also called Dyer's Greenwood.

This is why it is so important to identify every plant by its botanical name. Also under this heading, I have endeavored to give the Spanish, French, German, and sometimes the Italian and Russian names as well.

**Life Span** A hardy annual (H.A.) is a plant raised from seed, which flowers and dies within the year.

A half-hardy annual (H.H.A.) can stand no frost,

and is usually started indoors. It is not planted out until all danger of frost is past. It will usually die with the first frost of fall.

A biennial (B) is sown one year and produces flowers and seed the following year, before it dies. Sometimes the length of life of a biennial may be prolonged by removing flower heads as they form.

A hardy perennial (H.P.) lives for at least two years and sometimes for as long as 20. Some stay green all winter, others die back and reappear in spring.

**Habitat** The natural home of the plant.

**Appearance** A description of the plant.

**Cultural Requirements and Propagation** Type of soil required, whether the situation should be dry or moist; whether good drainage is important. Whether sun, shade, or partial shade suits the plants best. How far apart they should stand. What pruning is necessary, and any special care needed during growth. How the plant can be propagated, viz. by seed, cuttings, offshoots, layering or division.

For details of these, and many more methods of

propagation, see "Handbook of Propagation", a Brooklyn Botanic Garden Record, Plants and Gardens, Volume 13, No. 2. Price $1.50.

**To Dye**

*Harvest* — Optimum time to dig or cut, for use in dyeing.

*Use* — Whether the material should be used fresh or dry.

*Part to use* — Whether the flowers, leaves, stems, roots, berries, or bark, or a combination of two or more of these should be used to obtain the dye.

*Material*
*Mordant*
*Color*
*Fastness* — The information under these four headings will indicate to you what materials may be dyed with this plant, what mordant is needed to give which colors, and how fast those colors should be. It should be noted that:

1. Color varies greatly with plants grown in different areas, due to mineral content of the soil, and various other factors of growth.
2. The heading "Fastness" refers to the fastness of the dye to light and washing. If the dye is fast to one, and not the other, it will be indicated.

*Note* — Various points of interest not covered under any of the above headings
Reference to a particular recipe, in one of the books listed in the bibliography.

# Part 1

# How to Grow 23 Herbs and 7 Plants for Dyeing

AGRIMONY—*Agrimonia eupatoria*

# AGRIMONY

| | |
|---|---|
| **Botanical Name** | Agrimonia eupatoria |
| **Botanical Family** | Rosaceae |
| **Other Names** | Sticklewort, Church Steeples, Cockleburr (Sp.) Agrimonia (Fr.) Herniole (Ger.) Bruchkraut (It.) Erniaria (Rus.) Repeiki |
| **Life Span** | Hardy perennial |
| **Habitat** | Most of Europe and England, and northern and eastern United States. Similar species of eupatoria are found all over the temperate Northern Hemisphere. |
| **Appearance** | Growing to a height of two or three feet (60 to 90 cm.), the rough stems are almost branchless. The leaves near the ground are as long as eight inches (20 cm.), decreasing in size as they go up the stem, and are much divided and saw-toothed. The sulphurous yellow flowers are small but numerous, closely arranged on tall spikes. These sweet scented flowers bloom from early summer to early fall. The seed pods are burr-like and bristly. The roots are long, black and woody. |
| **Cultural Requirements** | A well drained ordinary soil with some shade. Young plants should be planted in clumps of six or eight, with six inches (15 cm.) between each one, and three feet (90 cm.) between the clumps. Agrimonia odorata can also be used for dyeing, |

but needs a more acid soil and full shade.

**Propagation** — By seed, when fresh in the fall, though germination is poor. By root cuttings in fall, or as the first shoots begin to show in spring.

**To Dye**

*Harvest* — The best yellows will be obtained from late fall harvested plants; gathered earlier they will yield a yellowish-buff color.

| | | |
|---|---|---|
| *Use* | Fresh | Fresh |
| *Part to use* | Leaves & stems | Leaves & stems |
| *Material* | Wool and cotton | Wool |
| *Mordant* | Alum | Chrome |
| *Color* | Brassy yellow | Gold |
| *Fastness* | Very good | Very good |

*Note* — An English dyer got orange from plants gathered in west Devonshire; a Canadian from Vancouver Island got orange (sometimes!) from mordanting with tin, and somebody else claims a dark green with an iron mordant. So try YOUR luck!

# BARBERRY

| | |
|---|---|
| **Botanical Name** | Berberis vulgaris |
| **Botanical Family** | Berberidaceae |
| **Other Names** | Berbery, Pipperidge bush. (Sp.) Agracejo (Fr.) Epinevinette (Ger.) Die Berberitz/Sauerdorn (Rus.) Nezreli Vinograd |
| **Life Span** | Hardy perennial |
| **Habitat** | Much of Europe, some parts of England, northern Africa and temperate Asia. Has become naturalized in the eastern United States. Berberis canadensis is indigenous to much of the United States, and more temperate parts of Canada; it is similar, though grows less high. |
| **Appearance** | A bushy deciduous shrub growing ten feet (3 meters) high. The leaves are small, tapering into a small foot-stalk. The flowers, appearing in late spring, are small, pale yellow and pendulous. The berries are red and acid. |
| **Cultural Requirements** | It is not demanding, and will thrive in ordinary well drained soil, in sun or shade. No pruning is required. |
| **Propagation** | By fresh seed in fall, in a sheltered spot. By cuttings of firm shoots in sandy soil in fall. By layering in fall; these layered branches should be left undisturbed for 12 months before planting |

BARBERRY—*Berberis vulgaris*

out into their permanent positions, four feet (1.20 meters) apart.

**To Dye**

| | | |
|---|---|---|
| *Harvest* | Bark and roots should be harvested in late summer and fall. Leafy tips can be used, but the colors are not strong. | |
| *Use* | Fresh or dried | Fresh or dried |
| *Part to use* | Inner bark | Roots and bark |
| *Material* | Wool, linen and cotton | Silk |
| *Mordant* | None | Tin (with acetic acid) |
| *Color* | Yellow | Yellow |
| *Fastness* | Fair to good | Good |
| *Note* | Alum can be used on wool, to give a lighter bright yellow. | |

The inner bark is suitable as a dye for leather. The berries are edible, and make good preserves.

BEDSTRAW—*Galium verum*

# BEDSTRAW

| | |
|---|---|
| **Botanical Name** | Galium verum |
| **Botanical Family** | Rubiaceae |
| **Other Names** | Our Lady's Bedstraw, Yellow Bedstraw (Sp.) Galio verdadero (Fr.) Gaillet jaune (Ger.) Wahres Lapkraut (It.) Gaglio giallo |
| **Life Span** | Hardy perennial, withstanding freezing temperatures |
| **Habitat** | Most of western Europe, coastal areas of Great Britain, and the eastern United States. Cultivated in the Hebrides. |
| **Appearance** | Growing to a height of one to three feet (30 to 90 cm.), it's stems are square and spindly. Its tiny leaves are fine and numerous, growing close together, and radiating from the stem like the spokes of a wheel. The yellow flowers are closely clustered at the tops of the stems, and bloom from mid to late summer. The whole plant spreads quickly by underground running roots. |
| **Cultural Requirements** | A light well drained soil, which has been manured and well worked, gives good results. When the plants reach a height of one foot (30 cm.) they should be staked and tied. Plant eight to twelve inches apart (20 to 30 cm.). In a cool, temperate climate, it needs full sun. |

**Propagation** By fresh seed in fall, or by root division in early spring. The stems can be layered in early summer.

**To Dye**

| | | | |
|---|---|---|---|
| *Harvest* | Dig roots in fall | Dig roots in fall | Dig roots in fall |
| *Use* | Fresh or dried | Fresh or dried | Fresh or dried |
| *Part to use* | Roots* | Roots* | Roots* |
| *Material* | Wool | Wool | Wool |
| *Mordant* | Alum | Chrome | Iron |
| *Color* | Red | Purplish/red | Plum |
| *Fastness* | Good | Good | Good |

*Note* *Roots should be boiled for about two hours to extract dye color. The fresh tops of plants yield a dull yellow dye when mordanted with alum.

**Of General Interest** Another name for Bedstraw was Cheese Rennet, from its ability to curdle milk. At the same time it gave the cheese a distinctive color. It was also used to stuff mattresses, and according to medieval legend was used by "Our Lady" at Christ's birth.

# BLOODROOT

| | |
|---|---|
| **Botanical Name** | Sanguinaria canadensis |
| **Botanical Family** | Papaveraceae |
| **Other Names** | Indian Paint, Red Puccoon, Tetterwort, Snakebite, etc. (Sp.) Sanguinaria (Fr.) Sanguinaire (Ger.) Blutwurzel |
| **Life Span** | Hardy perennial |
| **Habitat** | In open woodlands in many parts of Canada, and south through the United States, to northern Florida and west as far as Arkansas. |
| **Appearance** | Growing to about six inches (15 cm.) in height, this erect little plant has long, downy-green leaves, which appear wrapped around the flower buds. The flowers are white, with yellow stamens, blooming from very early spring to mid-summer. The roots are thick and contain an orange-red juice. |
| **Cultural Requirements** | Prefers a sandy loam or peat in a well drained position, with filtered sunlight and some total shade. The plants should be watered freely in dry weather, and mulched in late winter with compost or leaves. If possible, they should be planted amongst maidenhair ferns, to give shade and prevent them drying out. In any case, the rhizomes should be planted six inches (15 cm.) apart, and about one inch (2½ cm.) deep. |

BLOODROOT—*Sanguinaria canadensis*

**Propagation**

By seed sown in a mixture of equal parts of leaf-compost, peat and sand, in spring or early fall, where some protection can be given. Seedlings may be transplanted to permanent situation when large enough to handle, and when weather has settled. Roots may also be divided in early spring or fall; the latter is better when digging for use in dyeing.

**To Dye**

*Harvest*

The roots should be harvested in fall, after the leaves die down. They should be stored in a dry place, or they will deteriorate quickly.

| | | | |
|---|---|---|---|
| *Use* | Fresh | Fresh | Fresh |
| *Part to use* | Roots | Roots | Roots |
| *Material* | Wool & silk | Wool & silk | Wool & silk |
| *Mordant* | None | Alum/Cream of tartar | Tin |
| *Color* | Orange | Rust | Reddish/pink |
| *Fastness* | Good | Good | Good |

*Note*

1. "Simultaneous" dyeing is recommended. Only 8 oz. (240 g.) of root is needed to 4½ US gallons (18 liters) of water. Soak roots for one hour, then boil for 30 minutes, strain, and use.
2. Mincemoyer Nursery of New Jersey list Bloodroot.

BROOM—*Cytisus scoparius*

# BROOM

| | |
|---|---|
| **Botanical Name** | Cytisus scoparius |
| **Botanical Family** | Leguminosae |
| **Other Names** | Scotch or European Broom, Bisom, Bizzom, Genista scoparius (Sp.) Citiso Europea |
| **Life Span** | Hardy perennial |
| **Habitat** | Many parts of England, temperate Europe and southern Asia. Also in temperate parts of the central United States and the more temperate parts of Canada. |
| **Appearance** | A bushy shrub growing to a height of five feet (1½ meters) or more. The small leaves are carried on slender, green flexible stems. The butterfly-shaped flowers are bright yellow, and bloom from late spring to early summer. They are followed by oblong black pods, which turn nearly black when ripe, and burst, sending their seeds in all directions. (Note: remove pods before ripe, or seedlings will appear all over the garden.) |
| **Cultural Requirements** | A dry sandy soil, with good drainage, in full or partial shade. Pruning is done when the flower shoots are cut for dyeing; if this pruning is to be heavy, twenty four inches (60 cm.) is sufficient space between plants. See note above. |

**Propagation** By seed in early spring. By cuttings in early fall, in a sheltered place.

**To Dye**

*Harvest* When flowering, the top 12 to 18 inches (30 to 45 cm.), should be cut. Do not cut back into the old wood.

| | | | |
|---|---|---|---|
| *Use* | Fresh | Fresh | Fresh |
| *Part to use* | Flowering tops and young shoots | Flowers | Flowers |
| *Material* | Wool | Wool | Wool |
| *Mordant* | Alum | Alum | Chrome |
| *Color* | Bright greenish/yellow | Bright yellow | Deep yellow |
| *Fastness* | Good | Good | Good |

*Note* The "simultaneous" method of dyeing is recommended. This plant will top dye on blue, giving good greens.

# CALLIOPSIS/COREOPSIS

| | |
|---|---|
| **Botanical Name** | Calliopsis tinctoria |
| **Botanical Family** | Compositae |
| **Other Names** | Tick seed, Yellow dye flower (Fr.) Coreopsis |
| **Life Span** | Half-hardy annual |
| **Habitat** | Grows in temperate climates in most parts of the world |
| **Appearance** | The plant grows to a height of 20 inches (50 cm.) or more. The stems are flexible, and the fine leaves insignificant. The flowers, blooming from mid-summer till first frost, have yellow ray-like petals and dark brown centres. |
| **Cultural Requirements** | Plant in ordinary soil, in a sunny well drained position. Thin seedlings to six inches (15 cm.) apart. They do not transplant successfully. |
| **Propagation** | By seed in late spring, where plants are to flower. |

**To Dye**

| | | |
|---|---|---|
| *Harvest* | When flowers are in full bloom | |
| *Use* | Fresh | Fresh |
| *Part to use* | Flowers | Flowers |
| *Material* | Wool | Wool |
| *Mordant* | Chrome | Tin |
| *Color* | Red/orange | Yellow |
| *Fastness* | Good | Good |

CALLIOPSIS/COREOPSIS—*Calliopsis tinctoria*

*Note*

The "simultaneous" method of dyeing is best. Some authorities state that cotton cannot be dyed with this plant. However, very full details for dyeing cotton with Calliopsis flowers will be found in Brooklyn Botanic Garden "Natural Plant Dyeing" handbook, Volume 29, No. 2, page 33, article by Esther Hasel. She also states that dried flowers may be used, and keep for years.

Calliopsis refers to the annual varieties and Coreopsis to the biennial or perennial varieties. Most varieties give a good dye.

DAHLIA—*Dahlia variablis*

# DAHLIA

| | | |
|---|---|---|
| **Botanical Name** | Dahlia variablis | |
| **Botanical Family** | Compositae | |
| **Other Names** | Georgina (Sp.) Dalia (Fr.) Dahlia | |
| **Life Span** | Half-hardy tuberous perennials | |
| **Appearance** | The height varies according to variety, from one to five feet (30 cm. to 1.50 meters). The flowers come in many forms and colors. | |
| **Cultural Requirements** | Soil well enriched with manure or compost, in a well drained sunny position. Tubers should be planted three inches (7½ cm.) deep, in late spring or early summer. Where winters are severe, tubers should be dug in fall, when leaves are dying down, and stored in a frost free place. Where winters are mild, they may be left in the ground. When grown from seed, the seedlings should be thinned to six inches (15 cm.) apart. | |
| **Propagation** | By tubers or by seed | |
| **To Dye** | | |
| *Harvest* | When flowers are in full bloom. Even when flowers have been in the house for several days, they may still be used for dyeing, with equally good results. | |
| *Use* | Fresh or dried | Fresh or dried |
| *Part to use* | Flowers | Flowers |

| | | |
|---|---|---|
| *Material* | Wool | Wool |
| *Mordant* | Chrome | Alum |
| *Color* | Clear orange | Brassy yellow |
| *Fastness* | Very good | Fair |

*Note*

1. Some authorities state that yellow flowers give clearer and brighter colors than pink ones. Others only that the color varies according to the color of the petals, from clear yellows to bright orange.
2. Tubers can be obtained from good garden shops in spring, or from dahlia growers direct. Seeds can be obtained from any good flower seed house.

# DYER'S BROOM

| | |
|---|---|
| **Botanical Name** | Genista tinctoria |
| **Botanical Family** | Leguminosae |
| **Other Names** | Dyer's Greenwood, Dyer's Weed, Woad-waxen, Wood-waxen. (Sp.) Retama (Fr.) Genet de Teintuiers (Ger.) Färbeginster (It.) Ginestra (Rus.) Drok |
| **Life Span** | Hardy perennial |
| **Habitat** | Common in many parts of England, but rare in Scotland. It is wild throughout Europe, and has established itself in the eastern United States, especially in New England. Also found in many parts of Russia. |
| **Appearance** | Growing to a height of 12 to 24 inches (30 to 60 cm.), the branching stems are stiff and spineless. The small narrow leaves are almost stemless. The bright yellow pea-like flowers bloom from mid-summer to early fall. The seed pods are brown, and about one inch (2½ cm.) long. |
| **Cultural Requirements** | Ordinary soil with good drainage, in the full sun. Plant 24 inches (60 cm.) apart, from fall to spring. To obtain a profusion of bloom, prune back flowers as they appear. |
| **Propagation** | By seed in spring, first soaking seed for 24 hours in water at room temperature. By layering in the fall. |

DYER'S BROOM—*Genista tinctoria*

**To Dye**

| | | |
|---|---|---|
| *Harvest* | When plants are in flower | When plants are in flower |
| *Use* | Fresh or dried | Fresh or dried |
| *Part to use* | Flowering tops, small leaves and stems | Flowering tops, small leaves and stems |
| *Material* | Wool | Wool |
| *Mordant* | Alum | Chrome |
| *Color* | Light yellow | Deep yellow |
| *Fastness* | Excellent | Excellent |
| *Note* | This plant may also be used to dye linen and leather. It is also an excellent top dye on blues, to give greens. | |

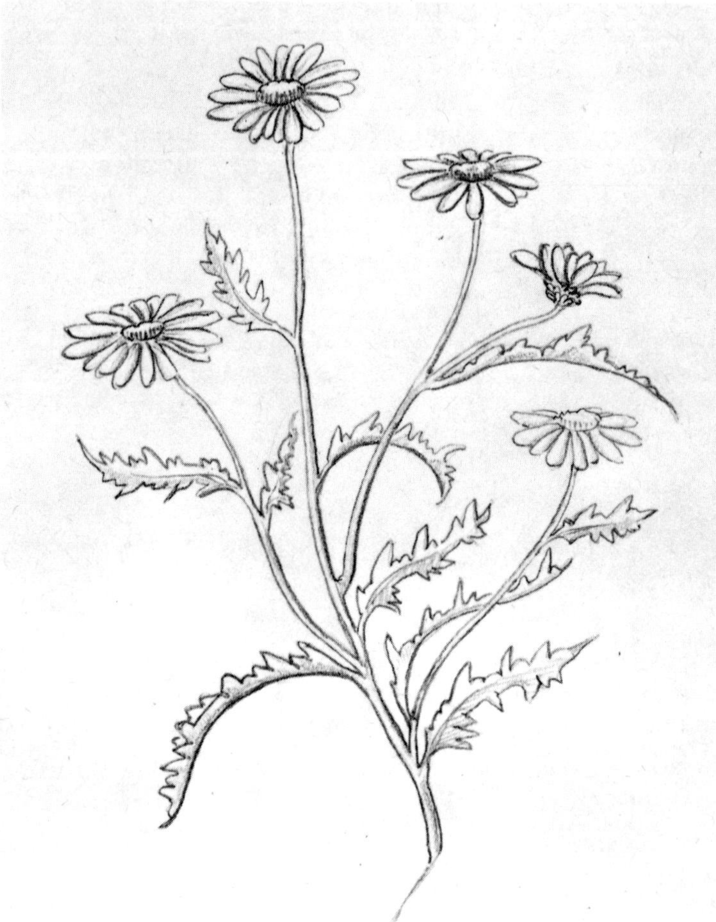

DYER'S CHAMOMILE—*Anthemis tinctoria*

# DYER'S CHAMOMILE

| | |
|---|---|
| **Botanical Name** | Anthemis tinctoria |
| **Botanical Family** | Compositae |
| **Other Names** | Golden Marguerite, Oxeye Chamomile |
| **Life Span** | Perennial, but winter killed by heavy frosts |
| **Habitat** | Not known |
| **Appearance** | Growing to a height of 24 inches (60 cm.), with stiff stems. Its foliage is fine cut, and the golden-yellow daisy like flowers bloom freely from late summer to early fall. |
| **Cultural Requirements** | A dry, sandy, well drained soil in full sun. Once established, these plants need little attention. Plant 10 inches (25 cm.) apart. |
| **Propagation** | By seed sown in spring. Germination takes up to three weeks in an optimum temperature of 70°F (21° C). Cuttings, taken in summer, should be over-wintered under glass, and planted out when there is no risk of frost. |

**To Dye**

| | | | |
|---|---|---|---|
| *Harvest* | In full bloom | | |
| *Use* | Fresh or dried | Fresh or dried | Fresh or dried |
| *Part to use* | Flowers | Flowers | Flowers |
| *Material* | Wool | Wool | Wool |
| *Mordant* | Alum | Chrome | Tin |
| *Color* | Yellow/buff | Golden/orange | Clear yellow |
| *Fastness* | Fair to good | Excellent | Excellent |
| *Note* | Good dyeing instructions will be found in "Home Dyeing with Natural Dyes" pages 14 and 15, and in the Brooklyn Botanic Garden's "Dye Plants and Dyeing" handbook, Volume 20, No. 3, page 22. | | |

# ELDER

| | |
|---|---|
| **Botanical Name** | Sambucus nigra (Europe) Sambucus canadensis (North America) |
| **Botanical Family** | Caprifoliaceae |
| **Other Names** | Dane-wort, Pio tree (Sp.) Sauco (Fr.) Sureau (Ger.) Hollunder (Anglo-Saxon) Eldrum |
| **Life Span** | Hardy perennial |
| **Habitat** | S. nigra is indigenous to Great Britain, much of Europe, Turkey and North Africa, also to parts of western Siberia. S. canadensis is indigenous to Canada, from New Brunswick to British Columbia; south into the United States as far as Texas, and to Colorado, Utah and Arizona. |
| **Appearance** | S. nigra will grow to a height of 20 feet (about 6 meters), but it is not often seen as tall as this, as it grows in hedges in its wild state, and is kept cut back, causing a more shrubby growth. Its pithy stem is pearly grey and rough. The deep green leaves turn purplish in the fall before they drop. The creamy-white, flat topped masses of blossom appear in early summer, and are followed by the black berries in early fall. S. canadensis is similar in appearance, but is more shrubby by nature, growing from five to ten feet (1.50 to 3 meters) high. |

ELDER—*Sambucus nigra*

**Cultural Requirements**

Tolerates an ordinary soil, in a moist situation in sun or partial shade. Once planted, they can remain undisturbed indefinitely. S. canadensis requires a damper, richer soil. Plant two feet (60 cm.) apart.

**Propagation**

Cuttings may be taken from fall to early spring, and inserted in moist soil out of doors. Cuttings will need about two to three years growth before they are available for use. From seed, sown as soon as ripe in fall; plants will need at least four years growth before they can be used.

**To Dye**

| | | | |
|---|---|---|---|
| *Harvest* | Berries when fully ripe | | |
| *Use* | Fresh or frozen | Fresh or frozen | Fresh or frozen |
| *Part to use* | Berries | Berries | Berries** |
| *Material* | Wool/silk | Wool/silk | Wool/silk |
| *Mordant* | Alum | Tin | Chrome |
| *Color* | Violet* | Blue/grey | Blue |
| *Fastness* | Fair to good | Fair to good | Fair to good |

| | | |
|---|---|---|
| *Harvest* | Leaves and bark as needed | |
| *Use* | Fresh | Fresh |
| *Part to use* | Leaves (young) | Older bark |
| *Material* | Wool | Wool |
| *Mordant* | Alum | Iron |
| *Color* | Yellow/lime | Grey/black |
| *Fastness* | Good | Good |

*Note*

*A lilac color can be obtained by using salt in the dye bath.

**A greenish/grey color can be obtained by adding pokeberries to the elderberries. To obtain dark grey with elderberries top dye the blue/grey with walnut.

**Of General Interest**

Mrs. Greive in her "A Modern Herbal," says that the English summer does not arrive until the elder is in full bloom, and ends when the berries are ripe. Ten pages of information follow, including the history and folklore, and medicinal, kitchen and dyeing recipes.

GOLDEN ROD—*Solidago canadensis*

# GOLDEN ROD

| | |
|---|---|
| **Botanical Name** | Solidago canadensis |
| **Botanical Family** | Compositae |
| **Other Names** | Woundwort, Aaron's Rod (Sp.) Barra de Oro (Fr.) Verge d'Or (Ger.) die Goldrute |
| **Life Span** | Hardy perennial |
| **Habitat** | Great Britain, much of Europe, central Asia, temperate parts of North America |
| **Appearance** | Growing two to three feet (60 to 90 cm.) in height, with stiff stems, and bright green, long alternate leaves. The flowers spring from a common central stalk, and terminate in a cluster of brilliant golden blossoms. |
| **Cultural Requirements** | The soil should be rich, moist but well drained. Plants will thrive in sun or shade. Every three years they should be lifted, divided, and replanted in the fall, in groups of five, leaving three feet (90 cm.) between groups. |
| **Propagation** | By seed sown in spring. By division of roots, between early fall and early spring. |
| **To Dye** | |
| *Harvest* | The whole plant (except roots) from early summer to frost. The flowers as they begin to bloom. |

| | | | |
|---|---|---|---|
| *Use* | Fresh | Fresh | Fresh |
| *Part to use* | Plant | Flowers | Flowers |
| *Material* | Wool, linen and cotton | Wool & silk | Wool & silk |
| *Mordant* | Iron | Alum | Chrome |
| *Color* | Yellow/green | Yellow | Old gold |
| *Fastness* | Fair | Good | Good |

| | |
|---|---|
| *Use* | Fresh |
| *Part to use* | Flowers |
| *Material* | Wool and silk |
| *Mordant* | Tin |
| *Color* | Bright yellow |
| *Fastness* | Good |
| *Note* | "Home Dyeing with Natural Dyes" gives good instructions for using Golden Rod (page 19), and for top dyeing with Madder and Indigo (pages 34 and 35). Solidago virgaurea and many others of the Solidago species give a good dye. |

## *CORRECTIONS*

*On page 49 under Appearance the text should read:*

*Varying in height from 3" (7.5cm) to 36" (90cm) this evergreen shrub has small narrow leaves. The flowers appear in August and September in the axils of the leaves on the upper stems. They vary in color from white through pink to purple, depending on the variety.*

*It has been brought to the author's attention that the illustration on page 50 is not CALLUNA VULGARIS, but one of the ERICAS. She would like to apologize for this mistake. However the ERICAS too can be used for dyeing.*

# HEATHER

| | |
|---|---|
| **Botanical Name** | Calluna vulgaris |
| **Botanical Family** | Ericaceae |
| **Other Names** | Ling (Sp.) Brezo (Fr.) Bruyère |
| **Life Span** | Hardy perennial |
| **Habitat** | Great Britain, Scandinavia (though not found wild in North America, will grow where the climate is temperate, and the soil suitable). |
| **Appearance** | Growing to 12 inches (30 cm.) in height, this evergreen shrub often blooms in spring. The flowers are bell like and pinkish mauve. |
| **Cultural Requirements** | An acid soil well enriched with peat or leaf compost.<br>A soil completely free of lime. A moist situation, in full sun. Plants should be 18 inches apart (45 cm.). |
| **Propagation** | By division of plants in spring or fall, or by tip cuttings taken in early summer. |
| **To Dye** | |
| *Harvest* | As needed. Leafy tips can be dried for winter use, but better results are obtained with fresh Heather. |

HEATHER—*Calluna vulgaris*

| | | | |
|---|---|---|---|
| *Use* | Fresh | Fresh | Fresh |
| *Part to use* | Young green tips | Flower buds | Open flowers |
| *Material* | Wool | Wool | Wool |
| *Mordant* | Alum | Alum | Alum |
| *Color* | Yellow to lime | Gold | Light orange |
| *Fastness* | Good | Good | Good |

*Note* Expert opinions vary as to shade of colors obtained, due probably to soil conditions in which the heather was grown. There are many varieties of Calluna and Erica, which is called Heath, and by careful selection flowers can be had almost all the year round. The best dye is obtained from C. vulgaris.

HOLLYHOCK—*Althaea rosea*

# HOLLYHOCK

| | |
|---|---|
| **Botanical Name** | Althaea rosea |
| **Botanical Family** | Malvaceae |
| **Other Names** | (Sp.) Pasarosa (Fr.) Rose tremière (Ger.) Stockrose |
| **Life Span** | Hardy perennials and hardy annuals, both thriving in a temperate climate. |
| **Appearance** | Sometimes growing to a height of over eight feet (2½ meters). The leaves are widely spaced on the thick, tough stems. The flowers may be single or double rosettes, in shades of pink or red, or pure white. |
| **Cultural Requirements** | A rich soil, in a warm, sunny position sheltered from winds. Keep well watered. Mature plants should stand about three feet apart (90 cm.). |
| **Propagation** | By seed. The annual varieties should be planted in early spring. The perennial varieties should be planted in early summer, and will flower the following summer. |

**To Dye**

| | | | |
|---|---|---|---|
| *Harvest* | When the plants are in full bloom | | |
| *Use* | Fresh | Fresh | Fresh |
| *Part to use* | Blossoms | Red blossoms | Leaves |
| *Material* | Wool & silk | Wool | Wool |
| *Mordant* | Chrome | Alum | Alum |
| *Color* | Orange | Pink | Lime |
| *Fastness* | Good | Good | Good |

LILY OF THE VALLEY—*Convallaria majalis*

# LILY OF THE VALLEY

| | |
|---|---|
| **Botanical Name** | Convallaria majalis |
| **Botanical Family** | Liliaceae |
| **Other Names** | Our Lady's Tears, May Lily (Sp.) Lirio del Valle (Fr.) Muguet (Ger.) Maiblume |
| **Life Span** | Hardy perennial |
| **Habitat** | Temperate regions of the Northern Hemisphere |
| **Appearance** | The broad green leaves first appear in early spring, as a quill-like shoot, unfolding gradually to become two distinct leaves. The flowers grow on a thin stem, and turn from pale green when young to a snowy white when in full bloom, in late spring to early summer. The perfume is exquisite. The flowers are followed by tiny red berries. |
| **Cultural Requirements** | The soil should be a rich sandy loam, in a moist, shady situation. Beds should be dug and re-made every four or five years, and should be mulched in early winter. Liquid manure should be applied after flowering finishes, once a week until fall. |
| **Propagation** | By seed in early spring, but flowers may not appear for several years. By planting "pips" in late winter or early fall, which will flower the following year. By digging clumps of roots four |

inches square (10 cm.) and replanting them six inches (15 cm.) apart, in fall.

**To Dye**

| | | | |
|---|---|---|---|
| *Harvest* | From spring until the first frosts. The younger leaves give a greener yellow than the older ones. | | |
| *Use* | Fresh | Fresh | Fresh |
| *Part to use* | Leaves | Leaves | Leaves |
| *Material* | Wool & silk | Wool & silk | Wool & silk |
| *Mordant* | Alum | Chrome | Tin |
| *Color* | Yellow/pale green | Bronze | Bright yellow |
| *Fastness* | Fair to good | Fair to good | Fair to good |

# MADDER

| | |
|---|---|
| **Botanical Name** | Rubia tinctorum |
| **Botanical Family** | Rubiacae |
| **Other Names** | Robbia, Dyers' Madder (Fr.) Garance (Ger.) der Krapp |
| **Life Span** | Hardy perennial. Will stand freezing temperatures. |
| **Habitat** | The Mediterranean, central countries of Europe, and Turkey. It will also grow in southern England, much of France, Holland, and central Asia. See Note *2. |
| **Appearance** | The prickly stems, being weak, are procumbent, and will grow to eight feet (2.40 meters) in length. The long, narrow, rough leaves are carried in whorls. The flower shoots rise from the joints of the stems. The yellow flowers seldom appear before the third year, in early summer. They are followed by the tiny round seed pods. The roots are thick and fleshy, and reddish-brown in color. Side shoots run underground, before sending up shoots. |
| **Cultural Requirements** | A light, sandy well drained soil; preferably in a sunny position. Due to its habit of spreading by underground roots, plants should be given plenty of room, in a clean and weed free area. Plant 12 inches (30 cm.) apart. |

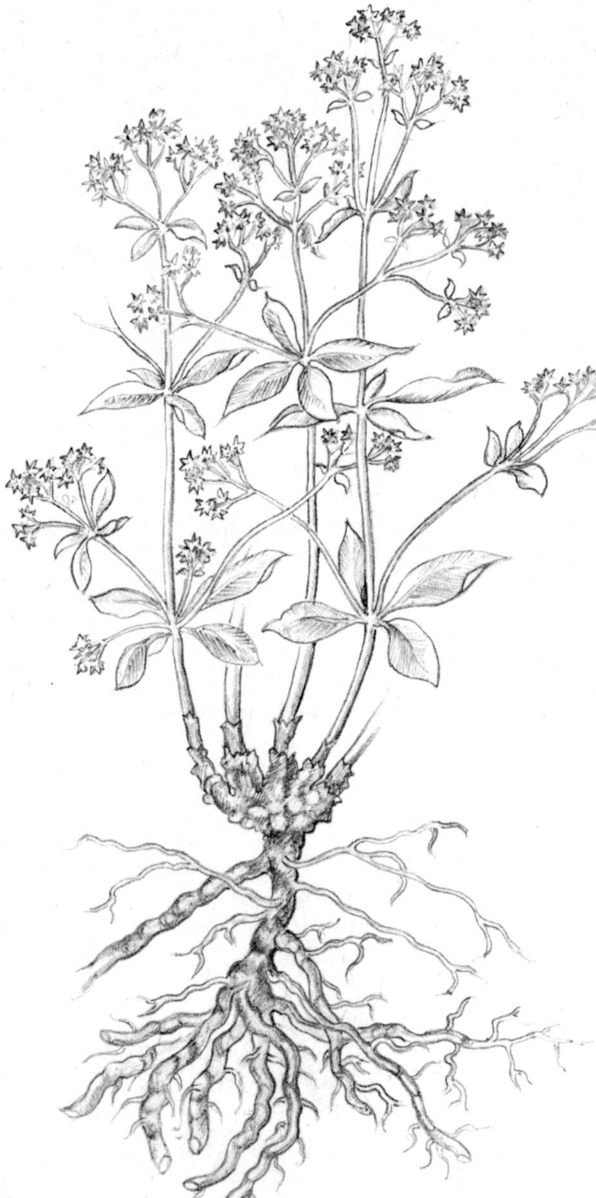

MADDER—*Rubia tinctorum*

| | |
|---|---|
| **Propagation** | By seed sown in spring; germination takes about ten days. By division of two or three year old roots in late spring or early fall. When digging to use for dyeing, pieces of root with a shoot can be replanted. |

**To Dye**

| | | | |
|---|---|---|---|
| *Harvest* | After three years of growth (if started from seed) or two years of growth (if started from plants), roots should be dug either before flowering in spring, or after the growth has died down in fall. | | |
| *Use* | Fresh and ground, or dried and powdered. | | |
| *Part to use* | Medium sized roots give better dye stuff than woody roots. | | |
| *Material* | Wool | Wool, cotton and silk | Wool |
| *Mordant* | Tin | Alum | Chrome |
| *Color* | Orange | Lacquer red | Garnet red |
| *Fastness* | Excellent | Excellent | Excellent |

*Special instructions for dyeing with Madder:*
Allow 8 oz. (225 g.) of Madder root to each pound (500 g.) of wool.
Using the "simultaneous" method of dyeing:

1. Put roots into a bag, and soak overnight in four and a half US gallons (18 liters) of water.
2. Enter wet, mordanted wool into the above bath.
3. Raise the temperature of the bath to 190° F(87° C) for alum mordanted wool, 140° F (60° C) for chrome mordanted wool, 212° F (100° C) for tin mordanted wool. BE CAREFUL TO DO THIS VERY SLOWLY, taking at least 90 minutes to reach the suggested temperatures. To keep the colors of the alum and chrome mordanted wool a clear red, do *not* exceed these temperatures.
4. Wool should be held at these temperatures for about 40 minutes.
5. Continue with the basic dyeing recipe (from No.12), page 119 but add ½ oz. (15 g.) of pure

soap flakes (not detergent) to the last rinse water.

*Note*

1. "Home Dyeing with Natural Dyes" (USDA) pages 25 and 26, and Robertson's "Dyes from Plants", page 94, give recipes for dyeing cotton with Madder.
2. *I have been able to find little information about growing Madder in North America, but I do know that it will grow in Kansas. My own original seed was generously sent me by a dyer, who was also a grower there. It will grow on southern Vancouver Island, Canada.
3. Fox Hill of Michigan and Taylors of California list Madder plants. I know of no commercial source of seed.

# MARIGOLD

| | |
|---|---|
| **Botanical Name** | Tagetes species |
| **Botanical Family** | Compositae |
| **Other Names** | French, African and Aztec Marigolds (Sp.) Tagetes or Copetes (Fr.) Souci |
| **Life Span** | Half hardy annual; enjoys high summer temperatures. |
| **Habitat** | Southern United States, Mexico, and central South America. |
| **Appearance** | Varying in height (according to the variety) from eight inches (20 cm.) to three feet (90 cm.). The leaves are fine cut, the flowers vary from pale brassy yellow to deep gold, and from copper to deep reddish orange. The plants flower profusely all summer till the first frosts. The whole plant has a strong odor, and has the reputation of discouraging nematodes in the soil. |
| **Cultural Requirements** | Well enriched soil in full sun. The seedlings should be hardened off by the time they are six weeks old. They should be planted out when all risk of frost is over, and the nights are beginning to warm up.* Place African Marigolds singly 12 inches (30 cm.) apart; French in groups of five at the same distance. |

MARIGOLD—*Tagetes species*

By seed in a temperature of 55 to 65° F (12 to 18° C).

**To Dye**

| | | |
|---|---|---|
| *Harvest* | When in full bloom | |
| *Use* | Fresh or dried | Fresh or dried |
| *Part to use* | Blossoms | Blossoms |
| *Material* | Wool or silk | Wool or silk |
| *Mordant* | Chrome | Alum |
| *Color* | Gold | Yellow-tan |
| *Fastness* | Good | Good |

*Note*

1. Recipes for dyeing cotton will be found in "Dyes from Plants" pages 87 to 89.
2. Plants may usually be obtained from garden shops in late spring and early summer. *Buy fairly late (see cultural requirements), and buy small plants which are not flowering, as they will settle down with less setback.

MEADOWSWEET—*Filipendula or Spirea ulmaria*

# MEADOWSWEET

| | |
|---|---|
| **Botanical Name** | Filipendula ulmaria or Spiraea ulmaria |
| **Botanical Family** | Rosaceae |
| **Other Names** | Queen of the Meadows (Sp.) Ulmaria, Reina de los Prados (Fr.) Ulmaire (Ger.) Wisenköenigin (It.) Reina dei Prati (Rus.) Luchicla |
| **Life Span** | Hardy perennial |
| **Habitat** | Parts of Great Britain, northern Europe, and Eurasia. Those found in the wild in the United States are probably garden escapes. |
| **Appearance** | Growing to a height of five feet (1.50 m.) or more. Its stems are reddish, leaves much toothed and downy. The massed heads of dainty, cream colored flowers appear in mid summer. The whole plant is aromatic. |
| **Cultural Requirements** | A rich, moist soil in partial shade. It must be kept well watered. Rhizomes should be lifted, divided and replanted every three or four years, spacing divisions 18 inches (45 cm.) apart. |
| **Propagation** | By seed in spring. By division of rhizomes in spring or fall. |
| **To Dye** | |
| *Harvest* | Stems and leaves as the plants bloom. Roots in late summer. |

| | | | |
|---|---|---|---|
| *Use* | Fresh | Fresh or dried | Fresh or dried |
| *Part to use* | Whole plant without roots | Roots | Roots |
| *Material* | Wool & silk | Wool & silk | Wool & silk |
| *Mordant* | Alum | Alum | Chrome |
| *Color* | Greenish yellow | Rosy red | Brown |
| *Fastness* | Fair | Fair | Fair |
| *Note* | The "simultaneous" method of dyeing is recommended when using plant tops. Black can be obtained with an iron mordant, but it is not too satisfactory. I have read that blue can be obtained with a mixture of Meadowsweet and Sorrel roots, but have never tried it. | | |

# MULLEIN

| | |
|---|---|
| **Botanical Name** | Verbascum thapsus |
| **Botanical Family** | Scrophulariaceae |
| **Other Names** | Velvet plant, Flannel leaf, Aaron's rod, etc. (Sp.) Gordoloro blanco (Fr.) Molene (Ger.) Wollkraut or Konigskerse |
| **Life Span** | Hardy biennial |
| **Habitat** | Throughout most of Great Britain and the temperate regions of Europe and Asia, where the soil is gravelly, sandy, or chalky. Also in areas of similar climate and soil in North America. |
| **Appearance** | During the first season of growth, a rosette of large, thick, hairy leaves appears. The second year, in spring, each plant puts up one sturdy, erect stalk often as high as seven feet (2 m.) high. The leaves are oblong and alternate and hug the stem. The flower spikes are thick, with small golden yellow flowers. |
| **Cultural Requirements** | Preferably gravelly or chalky soil, situated in sun or shade. Self-sown seedling should be transplanted when three or four leaves have formed, and spaced 18 inches (45 cm.) apart. |
| **Propagation** | Seed should be sown after all danger of frost has passed, in late spring or early summer. |

MULLEIN—*Verbascum thapsus*

**To Dye**

| | | | |
|---|---|---|---|
| *Harvest* | Late summer until first frost. | | |
| *Use* | Fresh or dry | Fresh or dry | Fresh or dry |
| *Part to use* | Leaves and stalks | Leaves and stalks | Leaves and stalks |
| *Material* | Wool & silk | Wool & silk | Wool & silk |
| *Mordant* | Alum | Chrome | Tin |
| *Color* | Yellow | Gold | Bright yellow |
| *Fastness* | Excellent | Excellent | Excellent |
| *Note* | Cotton and linen may also be dyed with Mullein leaves and stalks. | | |

ONION—*Allium cepa*

# ONION

| | |
|---|---|
| **Botanical Name** | Allium cepa* |
| **Botanical Family** | Liliaceae |
| **Other Names** | (Sp.) Cebolla (Fr.) Ognon (Ger.) Zwiebel (It.) Cipolla (Rus.) Luk |
| **Life Span** | Hardy perennial. Usually treated as an annual. |
| **Appearance** | As this is the common cooking onion, it needs no description. There are both red and yellow skinned varieties which may be used for dyeing. |
| **Cultural Requirements** | Deeply dug, well drained soil in full sun. Six to eight inches (15 to 20 cm.) space is required between mature onions. |
| **Propagation** | By seed sown in early spring or very late summer, if winters are not too severe. By "sets" planted in very early spring. |

**To Dye**

*Harvest* When bulbs are mature. Store in frost free place.

| | | | |
|---|---|---|---|
| *Use* | Fresh or dried | Fresh or dried | Fresh or dried |
| *Part to use* | Red skins | Red skins | Red skins |
| *Material* | Wool & silk | Wool & silk | Wool |
| *Mordant* | Tin | Chrome | Alum |
| *Color* | Tan/brown | Gold | Reddish orange |
| *Fastness* | Good | Good | Good |
| *Use* | Fresh or dried | Fresh or dried | Fresh or dried |

| | | | |
|---|---|---|---|
| *Part to use* | Red skin | Yellow skin | Yellow skin |
| *Material* | Cotton and linen | Wool and silk | Wool |
| *Mordant* | Chrome | Alum | Chrome |
| *Color* | Dark tan | Burnt orange | Brass |
| *Fastness* | Good | Good | Fair to good |

*Note* *The skins of the Potato onion (Allium cepa var. aggregatum) and the Shallot (allium ascolonicum) may also be used.

In "Dyes from Plants" Seonaid Robertson gives simple suggestions for the beginner, for dyeing with onion skins and no mordants. I obtained a great range of warm orange/browns using her recipes with the red skins of Potato onions.

# POKEWEED

| | |
|---|---|
| **Botanical Name** | Phytolacca decandra |
| **Botanical Family** | Phytolaccaceae |
| **Other Names** | Virginian Berry, Pokeroot, Pokeberry, Red ink plant, Pigeon-berry, Poke salad. (Sp.) Bella Sombra (Fr.) Mechoacan du Canada, Morelle a gràppes, Herbe de la laque (Ger.) Amerikanische Scharlachbeere, Kermesbeere. |
| **Life Span** | Hardy perennial |
| **Habitat** | Indigenous to most of North America. Also found in countries bordering on the north and south coasts of the Mediterranean Sea. |
| **Appearance** | Growing to a height of five feet (1.50 meters) or more. The stems are hollow and branching, the leaves ovate, turning from green to purple in the fall. The whitish flowers are followed by dense clusters of deep purple berries. The roots are large and branched. |
| **Cultural Requirements** | It will grow in any soil, in sun or shade, but prefers a damp situation. It is usually not until the second year after planting seed that the berries form. It can become a pestiferous weed. Birds will eat the berries, and the seed will become widely scattered. Plants should be eight to ten inches (20 to 25 cm.) apart. |

POKEROOT—*Phytolacca decandra*

**Propagation** By seed sown in spring or fall. By pieces of root with shoot attached, in spring or fall.

**To Dye**

| | | | |
|---|---|---|---|
| *Harvest* | When berries are fully ripe | | |
| *Use* | Fresh | Dried | Fresh |
| *Part to use* | Berries | Berries | Berries |
| *Material* | Wool & silk | Wool & silk | Wool & silk |
| *Mordant* | Alum | Alum | Tin |
| *Color* | Reds | Browns | Red |
| *Fastness* | Fair | Fair | Fair |

*Note* Although the fastness is only fair, Pokeberries are well worth experimenting with, as they give a variety of interesting colors. "Simultaneous" dyeing is recommended.

PRIVET—*Ligustrum vulgare*

# PRIVET

| | |
|---|---|
| **Botanical Name** | Ligustrum vulgare |
| **Botanical Family** | Oleaceae |
| **Other Names** | Ligustrum (Fr.) Troène commun (Ger.) Ligüster |
| **Life Span** | Hardy perennial |
| **Habitat** | South eastern Europe, north Africa, and western Asia. It has become naturalized in parts of Great Britain and the eastern United States. |
| **Appearance** | Growing to a height of six to ten feet (1.80 to 3 meters), if left unpruned. Its lance shaped, dark green leaves are smooth. Its fragrant white flowers are borne in terminal clusters; the black berries last till late fall. |
| **Cultural Requirements** | A well drained soil in the sun or partial shade. Plants should be placed eight to ten inches (20 to 25 cm.) apart, to form a hedge. |
| **Propagation** | By seed (berries) sown in late fall in a nursery bed, and transplanted to their permanent position the following fall. By cuttings two to four inches (5 to 10 cm.) long in summer; these need full shade until well rooted. By cuttings eight to twelve inches (20 to 30 cm.) in fall. |

**To Dye**

| | |
|---|---|
| *Harvest* | As needed, from early summer to first frost. |

Clippings should be saved when pruning, and the leaves stripped from the stems for use.

| | | | |
|---|---|---|---|
| *Use* | Fresh | Fresh | Fresh |
| *Part to use* | Leaves | Leaves | Leaves |
| *Material* | Wool | Wool | Wool |
| *Mordant* | Alum | Chrome | Iron |
| *Color* | Bright yellow | Gold | Dark green* |
| *Fastness* | Very good | Very good | Very good |

*Note* Use basic recipe up to and including No. 9, page 118. But for Nos. 10 and 11, simmer for ten to fifteen minutes, then check frequently until the desired color is obtained.

*Iron may be added direct to dye bath of alum mordanted wool, to obtain the dark green.

There are a number of varieties of Ligustrum, all of which are suitable for dyeing. Particularly attractive is the Golden evergreen variety (L. ovalifolium aureum), suitable for window boxes as well as low hedges. This variety also gives exceptionally good yellows.

# RAGWORT

| | |
|---|---|
| **Botanical Name** | Senecio jacobaea |
| **Botanical Family** | Compositae |
| **Other Names** | Ragweed**, Staggerwort, Squaw-weed, Cankerwort. (Ger.) Kreuzkraut |
| **Life Span** | Hardy perennial |
| **Habitat** | Great Britain, much of Europe, Siberia, and northwest India. Though not indigenous to the United States, it is found in some eastern states. |
| **Appearance** | Growing to a height of two to three feet (60 to 90 cm.), the branching stem is deeply furrowed, the glossy leaves fine-cut and yellowish green. The flat topped clusters of yellow flowers appear in mid to late summer. They are aromatic. (This plant is a close relative of the greenhouse cineraria.) |
| **Cultural Requirements** | Will thrive in a poor soil, preferably in full sun, and needs little attention beyond weeding. Plant 12 inches (30 cm.) apart. |
| **Propagation** | By seed, as soon as ripe in the fall. This plant will become a pestiferous weed if not kept under control, so flower heads should be removed before they develop seed. Ragwort is poisonous to stock. |

RAGWORT—*Senecio jacobaea*

**To Dye**

| | | | |
|---|---|---|---|
| *Harvest* | Leaves and stems as needed. Flowers as they come into bloom. | | |
| *Use* | Fresh | Fresh | Fresh |
| *Part to use* | Stems and leaves | Flower heads | Flower heads |
| *Material* | Wool | Wool | Wool |
| *Mordant* | Alum | Alum | Chrome |
| *Color* | Green | Rich yellow* | Brownish yellow |
| *Fastness* | Fair to good | Good | Good |
| *Note* | *Following basic recipe, remove at No. 11 as soon as desired color has been obtained, or wool will become brown.<br>**Ragweed of the United States is not a Senecio, but Ambrosia artemisiaefolia. | | |

SAFFLOWER—*Carthamus tinctoria*

# SAFFLOWER

| | |
|---|---|
| **Botanical Name** | Carthamus tinctoria |
| **Botanical Family** | Compositae |
| **Other Names** | Dyer's saffron, Bastard saffron, Dyer's or distaff thistle, False Saffron. (Sp.) Cartamo, Alazor, Azafran falso (Fr.) Carthane (Ger.) die Saflor (India) Koosumbha (China) Hoang-tchi |
| **Life Span** | Hardy annual |
| **Habitat** | There seems to be some disagreement among the experts as to where the Safflower originated. However, it will grow well in any temperate climate, as well as Egypt, parts of China and India, and most of Europe. |
| **Appearance** | Growing to a height of two to three feet (60 to 90 cm.), the stiff, upright stems branch at the top. The oval shaped leaves are prickly. The thistle like flowers, first appearing in mid summer, are a deep orange yellow. The fruit, often incorrectly referred to as the seed, resembles pearly white shells. |
| **Cultural Requirements** | Ordinary, well drained soil in full sun. The seedlings should be thinned to ten inches (25 cm.) apart, and kept well cultivated. |
| **Propagation** | The seed, sown in late spring, in a temperature of 60°F (15°C) will germinate in about 14 days. |

**To Dye**

*Harvest* When the flowers are opening, on a dry day.

| | | | |
|---|---|---|---|
| *Use* | Fresh or dried | Fresh or dried | Fresh or dried |
| *Part to use* | Flowers | Flowers | Flowers |
| *Material* | Wool & silk | Wool | Wool & silk |
| *Mordant* | Alum | Iron | Tin |
| *Color* | Clear yellow | Brass | Rust |
| *Fastness* | Good | Good | Good |

*Note* On page 42 of Brooklyn Botanic Garden "Dye Plants and Dyeing" (Volume 20, No. 3), details are given for obtaining a red dye, by first washing out the yellow. The Process seems somewhat complicated. This red dye was much used in cosmetics at one time. All authorities agree that it is an excellent dye for silk.

# SAFFRON

| | |
|---|---|
| **Botanical Name** | Crocus sativus |
| **Botanical Family** | Iridaceae |
| **Other Names** | (Sp.) Azafran (Fr.) Safran (Ger.) Saffran (It.)Zafferano (Rus.) Schafran (Arabian) Zaffer (Hebrew) Karkom (Gk.) Krokos |
| **Life Span** | Hardy perennial, some protection needed in sub-zero winters. |
| **Habitat** | Eastern Mediterranean to Middle East. This plant is cultivated in England, southern France, Italy, Sicily, Hungary, Greece, Turkey, Iran, India and China; but the largest commercial crops are grown in Spain. |
| **Appearance** | Growing only three inches (7 to 8 cm.) high, the lilac flowers, appearing in early fall, have the typical crocus form. The leaves are fine and grasslike. The orange-scarlet stigmas rise out of the open flowers. The tiny brown corms are flattened, nutlike and shiny. |
| **Cultural Requirements** | Soil should be rich, well worked and well drained. Plant corms six inches (15 cm.) apart in full sun or part shade, but sheltered from winds in mid summer (late July, early August). Corms should be lifted every three or four years, and during their dormant period, divided and replanted. |

SAFFRON—*Crocus sativus*

**Propagation** By division of three to four year old corms, or by seed planted in spring; it will take at least three years to obtain plants from seed.

**To Dye**

*Harvest* In early fall the stigmas are plucked out of the open flowers, and dried between layers of paper.

*Use* Dried

*Part to use* Stigma

*Material* Wool and silk

*Mordant* Alum

*Color* Yellow

*Fastness* Very good

*Note*

1. Saffron should be boiled gently for half an hour, then wool should be simmered until the required color is obtained. Silk should be steeped at about 160° F (70° C). Authorities vary in their estimates of the amount of flowers needed to produce a specific amount of Saffron. But harvesting is laborious, as many as 200,000 stigmas being needed to make one kilogram of Saffron (about 88,000 to the pound)!
2. Well-Sweep Herb Farm of New Jersey (see Appendix "A" for address) have bulbs for sale in August — and so have Herbs n' Honey Nursery of Oregon.
3. Autumn crocus (Crocus autumnale) is NOT Saffron.

TANSY—*Tanacetum vulgare*

# TANSY

| | |
|---|---|
| **Botanical Name** | Tanacetum vulgaris |
| **Botanical Family** | Compositae |
| **Other Names** | Bitter Buttons, Arbor Viteae, Parsley fern. (Sp.) Tanaceto, Yerba de San Marco (Fr.) Tanaise (Ger.) Rainfarn (It.) Tanaceto (Rus.) Ryabinska |
| **Life Span** | Hardy perennial |
| **Habitat** | Great Britain, and much of northern Europe. Although found wild in eastern North America, it is not indigenous, but has become naturalized. |
| **Appearance** | Growing to a height of two to three feet (60 to 90 cm.). The stems are erect, the oblong leaves deeply cut and fernlike. Its flowers, growing in clusters, are round, flat and buttonlike. It is strongly aromatic. Its roots creeping. |
| **Cultural Requirements** | It will thrive in any soil where the drainage is good, and it is in full sun. Two to three feet (60 to 90 cm.) should be left between plants, to allow for its spreading habit. Plants should be dug and divided every three or four years. |
| **Propagation** | By seed in spring. By division of roots, or "slips" with a piece of root attached, in spring or fall. |

**To Dye**

| | | |
|---|---|---|
| *Harvest* | Leaves before flower heads appear. Flower heads as they open. | |
| *Use* | Fresh | Fresh |
| *Part to use* | Young leaves | Flowers |
| *Material* | Wool | Wool |
| *Mordant* | Alum | Alum |
| *Color* | Yellowish green | Greenish yellow |
| *Fastness* | Fair | Fair |

# TOMATO

| | |
|---|---|
| **Botanical Name** | Lycopersicum esculenteum |
| **Botanical Family** | Solanaceae |
| **Other Names** | Love apple (Sp.) Tomate (Fr.) Pomme d'Amor (Ger.) Liebesaepfel |
| **Life Span** | Half hardy annual |
| **Appearance** | As this is the common garden tomato, it needs no description. Any variety may be used, miniature, staking or bush. |
| **Cultural Requirements and Propagation** | A well drained soil, which has been well limed and composted, in full sun. The seed should be started at a temperature of 65° F (18° C), the date depending on the availability of glass protection and heat. Plants should not be put out until ALL risk of frost has passed and the nights are beginning to warm up a little.<br><br>Staking varieties should be planted out 20 inches (60 cm.) apart, and be trained to one stem, removing side shoots as they form. Bush varieties should be planted three feet (90 cm.) apart, and left unstaked and unpruned. |

TOMATO—*Lycopersicum esculentum*

**To Dye**

| | | |
|---|---|---|
| *Harvest* | Late summer and early fall | |
| *Use* | Fresh | Fresh |
| *Part to use* | Vine | Vine |
| *Material* | Wool & silk | Wool & silk |
| *Mordant* | None | Alum |
| *Color* | Reddish brown | Light yellow |
| *Fastness* | Good | Fair to good |

WELD—*Reseda luteola*

# WELD

| | |
|---|---|
| **Botanical Name** | Reseda luteola |
| **Botanical Family** | Resedacae |
| **Other Names** | Dyer's mignonette, Dyer's rocket. (Sp.) Reseda de teñir (Fr.) Réséda de Teinturiers (Ger.) der Wan |
| **Life Span** | Hardy annual |
| **Habitat** | In Great Britain, where soil is chalky. Mr. W. Robertson, in the Brooklyn Botanic Garden handbook, states that it has appeared in a Swiss neolithic settlement. It is also found throughout Mediterranean countries. |
| **Appearance** | Growing to a height of six feet (1.80 meters), or more, with dark green attractively crimped foliage. The inconspicuous greenish yellow flowers, appearing in early summer, are carried on long, erect spikes. |
| **Cultural Requirements** | A sandy, well limed, well drained soil, in sun or partial shade. The seedlings should be thinned to ten inches (25 cm.) apart, when small; they do not transplant successfully. |
| **Propagation** | By seed planted in early spring. This plant self sows very freely. |

**To Dye**

| | | | |
|---|---|---|---|
| *Harvest* | When in full flower, but well before blossoms begin to fall. | | |
| *Use* | Fresh or dried, (the dried flowers are considered by some dyers to give a less clear color). | | |
| *Part to use* | Leaves, flowers, and stalks. | | |
| *Material* | Wool & silk | Wool & silk | Wool & silk |
| *Mordant* | Alum | Chrome | Tin |
| *Color* | Lemon yellow | Golden yellow | Yellow orange |
| *Fastness* | Good | Good | Good |

| | |
|---|---|
| *Part to use* | Leaves, flowers, and stalks. |
| *Material* | Wool and silk |
| *Mordant* | Iron |
| *Color* | Olive green |
| *Fastness* | Good |
| *Note* | In basic recipe (at No. 5 on page 118 ):<br>5. Take two hours to bring to boil, and simmer for 45 minutes.<br>6. Strain etc., continue as in basic recipe.<br>"Dyes from Plants" gives recipes for dyeing cotton, on page 90. |

# WOAD

| | |
|---|---|
| **Botanical Name** | Isatis tinctoria |
| **Botanical Family** | Cruciferae |
| **Other Names** | (Sp.) Pastel (Fr.) Guède (Ger.) der (Faber) Waid (It.) Guado (Dutch) Weat (Anglo-Saxon) Wad |
| **Life Span** | Biennial |
| **Habitat** | Southern Europe and western Asia. It has become naturalized as far north as Sweden, in Europe, and parts of Great Britain; always where the soil is poor and stony. |
| **Appearance** | During the first year, a large rosette of bluish green leaves is formed. The second year, the flower stalks will reach a height of three to four feet (90 to 120 cm.). In early summer, the yellow flowers appear. The seeds are long and blue black. |
| **Cultural Requirements** | A rich, well cultivated soil is needed, but not recently manured. This will give better plants than those found in the wild. In full sun or partial shade. Each planting should be in a different area, and precautions taken against club root and cabbage fly (as is necessary for cabbages, cauliflowers, etc.). |
| **Propagation** | By seed in mid summer. Germination takes about 10 days. Seedlings should be thinned to eight inches (20 cm.) apart. |

WOAD—*Isatis tinctoria*

**To Dye**

| | |
|---|---|
| *Harvest* | Leaves can be removed as soon as they are mature, in the second year, until plant is coming into flower. |
| *Use* | Fresh |
| *Part to use* | Young leaves |
| *Material* | Wool |
| *Mordant* | None |
| *Color* | Blue, after exposure to air |
| *Fastness* | Fair to good |
| *Note* | To extract the dye from woad is a complicated business, according to most authorities, and interesting descriptions of the methods are described in the Brooklyn Botanic Garden handbook (Volume 20, No. 3) on page 34. Seonaid Robertson, in "Dyes from Plants" quotes extensively from William Rhind, and from a journal of the Royal Horticultural Society of 1901, and both sound too complicated for the average dyer to tackle. However, she also gives a "simple contemporary method", which is practical even if it results in a slightly less strong, and less fast dye, than the older methods. |

YELLOW FLAG —*Iris pseudacorus*

# YELLOW FLAG

| | |
|---|---|
| **Botanical Name** | Iris pseudacorus |
| **Botanical Family** | Iridaceae |
| **Other Names** | Yellow Iris, Dragon Flower, Iris aquaticus (Fr.) Fleur de Luce (Ger.) Wasser-schwertlilie |
| **Life Span** | Hardy perennial |
| **Habitat** | Great Britain, most of Europe, parts of Siberia, and north Africa. |
| **Appearance** | The sword-like leaves often grow to a height of four feet (1.20 meters) or more. The yellow flowers, appearing in summer, are carried on the ends of the stiff, erect stems. They belong to the "beardless" class of Iris. The thick, fleshy rhizomes creep horizontally under the surface of the soil. |
| **Cultural Requirements** | Ordinary moist soil, in part shade, preferably close to water. Beds should be dug up and replanted every three years, as soon as blooming is over. Rhizomes should be cut so that each piece has at least one bud, and replanted 18 inches (45 cm.) apart. |
| **Propagation** | By division of rhizomes in spring or fall. By seed sown in spring. Germination is difficult and slow. |

**To Dye**

| | |
|---|---|
| *Harvest* | When the plants have died down |
| *Use* | Fresh or dried |
| *Part to use* | Rhizomes |
| *Material* | Wool |
| *Mordant* | Iron |
| *Color* | Grey/black |
| *Fastness* | Good |
| *Note* | Color deepens from grey to black after simmering wool for several hours. |

# ZINNIA

| | |
|---|---|
| **Botanical Name** | Zinnia — various species |
| **Botanical Family** | Compositae |
| **Other Names** | Youth-and-old-age |
| **Life Span** | Half hardy annual |
| **Habitat** | Mexico |
| **Appearance** | The common garden Zinnias are so well known that they need no description. There are many varieties, of varying heights and colors, and all may be used for dyeing. |
| **Cultural Requirements** | A deep, loamy soil, well manured or composted, in a sunny situation. Keep well watered in dry weather. Mature plants need to be 12 inches (30 cm.) apart. |
| **Propagation** | For early blooms; by seed in a temperature of 55° F (13° C), indoors. The seedlings should be planted into individual pots, and hardened off gradually. They should not be planted out in the garden till all risk of frost is past, and the nights have begun to warm up.<br>For later bloom, direct planting out of doors, when all risk of frost has passed; the seed should be sown in groups of three, 12 inches (30 cm.) apart, and seedlings thinned to one in each group. |

ZINNIA—*Zinnia*

**To Dye**

| | | |
|---|---|---|
| *Harvest* | When flowers are in full bloom | |
| *Use* | Fresh or dried | Fresh or dried |
| *Part to use* | Flower heads | Flower heads |
| *Material* | Wool or silk | Wool or silk |
| *Mordant* | Alum | Chrome |
| *Color* | Bright yellow | Greenish yellow |
| *Fastness* | Fair to good | Fair to good |

# Part 2

# How to Use Herbs and Plants for Dyeing

# Equipment You Will Need To Mordant and Dye Your Wool

1. Soft water. If you cannot collect rain water, soften your tap water with a commercial water softener such as "Calgon". Follow the directions given on the packet. You can also soften it with acetic acid, or even with vinegar, testing with litmus paper to get the water "neutral".
2. Several enamel or stainless steel saucepans, large enough to hold four and a half US gallons (18 liters) of water AND your wool AND leave room to stir it all. If you are going to be handling less than one pound (500 g.) of wool at a time, relatively smaller pans will do.
3. Buckets and bowls for rinsing. They may be of plastic, as they do not have to be heated.
4. Standard plastic measuring spoons and glass measuring cups. Two and four cup sizes.
5. An accurate measure for one peck (a peck is a measure of capacity). A container which holds *two* US gallons, 256 fluid ounces, or (approximately) ten liters, holds one peck. Incidentally, there are four pecks to one bushel.
6. Fine strainers to remove plant material from liquid dye.
7. Smooth sticks for stirring; glass or plastic are best, as wooden ones must be changed after each different dye color.
8. Thermometer which registers up to at least 212 degrees Fahrenheit, or 100 degrees Centigrade or Celsius.
9. Scales accurate to ¼ oz. or 5 grams.
10. Rubber gloves.
11. Clothes line, in the shade.
12. Kitchen stove. If you can set up some means of heating the saucepans on a stove in a shed, or a shelter out of doors, it will save a lot of mess in the kitchen!

# Chemicals You Will Need To Mordant Your Wool

Wool (and other fibers) are mordanted to prevent fading and bleeding of the colors. According to which mordant is used, so is the shade and color determined. Many different chemicals are used to mordant wool, but to start, you need only worry about five.

1. Alum, or potassium aluminum sulphate (preferably NOT aluminum ammoniate, i.e. ammonium alum). Obtainable from builders' yards or dyers' supply houses.
2. Chrome, or potassium dichromate or sodium dichromate, also known as bichromate of potash. This is obtainable from a photographic supply shop, or dyers' supply houses. It is sensitive to light, and poisonous.
3. Cream of tartar, or tartaric acid, or potassium tartarate. Grocery stores have this, as do drug stores.
4. Tin, or stannous chloride, which can be obtained from chemical or dyers' supply houses, and sometimes from large pharmacies. It is poisonous.
5. Iron, or ferrous sulphate, also known as copperas or green vitriol. This can also be obtained from chemical or dyers' supply houses, and sometimes from large pharmacies.

# Get Everything Ready

1. Weigh wool to be dyed, while it is dry.
2. Prepare a soap solution ¾ oz. (20 g.) to 5 gallons (20 liters) of soft water. Wash and rinse waters should be at a temperature of 95°F (35°C).
3. Calculate the amount of chemicals needed to mordant the wool (see "How to Mordant").
4. Calculate the amount of plant material needed to dye the wool (see "How to Dye"). Always dye enough wool at the same time to complete any one project, as it is almost impossible to get exactly the same shade twice.
5. Make wool up into 8 oz. (240 gm) hanks or skeins for major projects. Tie up loosely with cotton thread. At the same time it is a good idea to also make up one, two and four ounce (30, 60 and 120 gms) hanks. These may then be taken from the dye pot at different times to obtain a variety of shades in each color.

# Some Dos and Don'ts When Handling Wool

1. When bringing liquids up to the boil (or near boil), do move the wool about in the bath, constantly and gently. Don't stir it round and round.
2. Do raise the temperature of the bath slowly. Don't put it over a high heat, to try and hurry things.
3. Do simmer the wool gently. Don't boil it.
4. Do keep the quantity of the water in the bath at the same level at all times, by adding more water at the same temperature as necessary.
5. Do make all changes of temperature as gradual as possible. Don't plunge wool from hot to cold liquids. For example, the first rinse of the dyed wool should be at the same temperature as the dye bath was. Gradually reduce the temperatures of the water in succeeding rinses, until the final one is no more than tepid.
6. Do squeeze the wool. Don't wring or twist it.
7. Do use a towel to remove excess water, after squeezing out all you can. Lay the wool on a towel on a flat surface, roll both of them into a "sausage", and press on the roll. Don't twist the towel and wool.
8. Do hang the wool in the shade to dry. Don't hang in the sun.
9. Do store in a cool, dry place. Don't store chrome mordanted wool in the light. It is better to use it immediately after mordanting.
10. Do leave the wool in the dye pot until it is a little darker than you require. Allow for it to be a little lighter in color when it has dried.

# Quantities

Remember that all weights and measures in the instructions are suggestions for handling one pound (or 500 grams) of wool; weighed, of course, when dry.

Metric Measures are not the equivalent measures, but the proportions are the same. This is done so that more convenient metric measures may be used. For example, one pound is equal to 448 grams . . . 500 grams actually equals 17.857 ounces, a little over one pound.

Smaller quantities may be used with perfect confidence, and measurements halved or quartered, according to how much wool is being used.

Always keep records of all your mordanting and dyeing, and samples of your finished product, with details attached.

1 pound of wool needs 4½ to 5 US gallons of liquid dye

or

500 grams of wool needs 18 to 20 liters of liquid dye.

It is difficult to generalize about what amount of dye plant material is needed per pound (or 500 grams) of wool. Until you gain some experience, you need some guidelines. Do not be afraid to increase the amounts of dye plant material. You are more likely to get good results with too much, than too little.

Allow 4½ to 5 US gallons (18 to 20 liters) of water, and:

One peck of flowers

or

One peck of leaves

or

One peck of stems (chopped and crushed)

or

One pound (500 grams) of berries

or

One half pound (250 grams) of roots

or

One and one half pounds (750 grams) of bark

to

EACH POUND (OR 500 GRAMS) OF WOOL

# How to Wash Wool

1. In the prepared soap solution, 95°F (35°C), gently squeeze the wool for several minutes.
2. Rinse thoroughly in water of the same temperature.
3. Repeat washing, using half the quantity of soap.
4. Rinse several times, to remove all traces of soap.
5. Remove from the water, and squeeze out as much as you can. DO NOT TWIST OR WRING.
6. Roll in a towel to remove any excess water.
7. Hang in the shade to dry, unless you are going to mordant it immediately.

If you want to start with a fleece, I would refer you to Seonaid Robertson's advice on page 12 of "Dyes from Plants". However, until you are an experienced dyer, I would recommend that you use purchased wool (be careful that it is 100% pure wool, and contains no synthetic fibers).

# How to Mordant Wool

Before mordanting any wool, it is necessary to soak it in cool water for a few minutes, unless it has just been washed, and is still wet. Squeeze it out gently before putting it in the prepared mordant. Always weigh the wool first, while it is dry.

## To Mordant With Alum

With 1 lb. or 500 g. of wool (weight when dry).
Use 4 oz. or 120 g. of alum.*
And 1 oz. or 30 g. of cream of tartar.
And 4 to 5 US gallons, or 16 to 20 liters of cold, soft water.
*If the wool to be mordanted is fine or delicate, use about 25% less alum. Too much will make the wool sticky and harsh.

1. Dissolve alum and cream of tartar in a little boiling water, and add it to the rest of the cold water. Stir well.
2. Start heating water, and when it is tepid, add the wet wool.
3. Continue heating slowly.
4. As water comes to the boil, lower heat.
5. Simmer gently for one hour.
6. Cool.
7. Leave wool in solution overnight.
8. Next day, remove wool from solution, squeeze gently, roll in a towel to remove surplus damp.
9. Wool may either be rinsed and dyed immediately, OR dried slowly and stored in a cool place, and then rinsed before dyeing at some other time.

## To Mordant With Chrome

It is advisable to mordant wool with chrome immediately before dyeing.

If you can't dye it immediately, keep it in a dark, cool place until you do.

With 1 lb. or 500 g. of wool (weight when dry).
Use ½ oz. or 15 g. of chrome.
And ¾ oz. or 20 g. of cream of tartar.
And 4 to 5 US gallons, or 16 to 20 liters of cold, soft water.

1. Dissolve the chrome and the cream of tartar in a little boiling water. Add this to the rest of the cold water, and stir well.
2. Start heating the water, and when it is tepid, add the wet wool.
3. Continue heating slowly, stirring now and then, BUT it is absolutely necessary to keep the wool covered, so as to exclude all light. Put a plate on top of the wool to keep it submerged; remove cover to stir only. Chrome is sensitive to light, and exposure of the chrome mordanted wool will cause uneven dyeing.
4. Simmer gently for three quarters to one and a half hours, depending on the thickness of the wool.
5. Continue to move wool about from time to time (don't stir it round and round).
6. Cool.
7. Squeeze gently, and roll in a towel to remove excess moisture.
8. Rinse in cool water to remove excess mordanting liquid.
9. Squeeze gently, and roll in towel to remove excess water.

## To Mordant With Tin

With 1 lb. or 500 g. of wool (weight when dry).
Use ½ oz. or 15 g. of tin.
And ½ oz. or 15 g. of cream of tartar.
And 4 to 5 US gallons, or 16 to 20 liters of cold, soft water.

1. Start heating water, and when tepid, add the cream of tartar, dissolved in a little boiling water, and stir well.
2. Add tin to the water (while still tepid) and stir again.
3. Add the wet wool.
4. Bring slowly to the boil.
5. Simmer for about an hour or more, depending on the coarseness of the wool. Move wool about continuously. Don't stir round and round.
6. Allow to cool, until it can be handled comfortably.
7. Remove from water.
8. Wash in soap and water of the same temperature.
9. Rinse at the same temperature.
10. Rinse again at a cooler temperature.
11. Rinse again in a slightly soapy water.

Sometimes tin is used as a mordant IN the dye bath, on alum mordanted wool. It is used towards the end of dyeing to brighten colors.

It is a difficult mordant to use, and tends to make the wool harsh and brittle. Extra care in washing and rinsing throughout all processes is necessary. A final soapy wash after dyeing tin mordanted fibers is recommended (see 11 above).

## To Mordant With Iron

Very little iron is needed, as it dulls colors; for this reason it is called a "saddener". It is usually used after the dyeing is finished, so is not truly a mordant.

With 1 lb. or 500 g. of wool (weight when dry).
Use ½ oz. or 15 g. of iron (less, for less dulling).
And 1 oz. or 30 g. of cream of tartar.

1. Dissolve the cream of tartar and the iron in a little boiling water.
2. Remove the dyed wool from dye bath, or hold it over the bath while the cream of tartar and iron solution is stirred in.
3. Return wool and stir; bring back to gentle simmer for 15 minutes.
4. Rinse thoroughly in two or three different waters.
5. Squeeze excess water out.
6. Roll in towel to dry further.
7. Dry in shade.

# How to Dye Wool

## A Basic Dyeing Recipe

1. Break and/or chop up the parts of the plant to be used.
2. If you are using roots and/or tough stems, thick leaves or bark, leave soaking overnight in two US gallons (8 liters) of cold, soft water.
3. In the morning, strain out the plant material (keeping the liquid), and put it into another two US gallons (8 liters) of cold, soft water.
4. If you are using berries, small thin leaves, or flowers, do *not* soak them, but put them into four US gallons (16 liters) of cold, soft water.
5. Bring either (3) or (4) to the boil slowly and simmer:
   Flowers for 10 to 20 minutes, or till they lose their color.
   Thin leaves for 30 minutes.
   Stems or berries for 45 minutes.
   Thick leaves or roots for 60 minutes or longer.
   Bark for one to two hours.
6. Strain out the plant material.
7. You will now have four US gallons (16 liters) of dye from the flowers and thin leaves.

   OR

   You will have two US gallons (8 liters) of dye from boiling the thick leaves, stems, roots and/or bark. This should be added to the dye obtained in the initial soaking period (3).
8. If the mordanted wool is dry, wet it thoroughly, and then squeeze out the surplus water.
9. Put the wet wool into the four US gallons (16 liters) of dye. Keep it moving in the water, and SLOWLY raise the temperature; 40 to 60 minutes is not too much time to allow it to reach a gentle simmer.
10. The time to simmer the wool in the dye bath varies. In general, dyes obtained from flowers need less simmering than those obtained from leaves; from 10 to 20 minutes as a rule. Though some flowers give a deeper color with longer simmering, others

"give" all their color in ten minutes. In general, an hour's simmering for dyes obtained from leaves, stems, and berries, and from one and a half to two hours for those obtained from roots and bark. During the dyeing period, if the water is getting low, lift out the wool, and add enough water (at the same temperature as the dye bath) to make up the original 4 US gallons (16 liters). Stir and return wool to the bath.

11. Continue simmering for the allotted time, or until you have obtained the color you want (see 10, page 111).
12. Allow to cool until the wool can be handled comfortably.
13. Transfer to a rinsing bath of the same temperature.
14. Rinse at least twice more, each successive rinse being cooler than the last.
15. Remove wool from water and squeeze it gently.
16. Dry, rolled in a towel.
17. Hang in the shade to dry.

NOTE: The dye bath is unlikely to be exhausted after dyeing one batch of wool. Several further batches of wool, mordanted with different mordants, can be dyed one after the other. Always dye iron mordanted wool last.

## Simultaneous Dyeing

This simply means that the initial step of extracting the dye in water is not done, but the chopped or crushed plant material is put into a thin cotton bag, and simmered in water with the wool.

Steps 2 to 7 (inclusive) are eliminated.

When this method is advisable, it is noted in the text, under the plant for which it is advised.

# Appendix "A"

This appendix lists names of growers of Herb seeds and plants mentioned in this book. You are unlikely to find all the seeds or all the plants for sale at any one source. Under each listing I have indicated if they sell either seeds or plants, or both. Each of them answered a questionaire, and many sent their catalogs to me. I apologize for any errors or omissions.

If you are unable to find the plants or seeds you want, listed under the common name by which you know them, look for them under their Latin name, or one of the names found under the heading "other names" in the text, pages 17 to 105. For example, in the catalogs Chamomile may be listed as Chamomile (Dyers)/ Anthemis tinctoria/ Golden Marguerite/ or Marguerite, Golden. Meadow Sweet as Filipendula ulmaria/ Spiraea ulmaria/ or Queen of the Meadows.

## CALIFORNIA

**J.L. Hudson, Seedsman — A World Seed Service,** P.O. Box 1058, Redwood City, CA 94064. Send 50 cents for their catalog. They sell Herb seeds wholesale and retail. Excellent listing of commonplace and hard to find varieties. Latin and common names given.

**Redwood City Seed Co.** (Craig Dremann) P.O. Box 361, Redwood City, CA 94064. Send 25 cents for their catalog. They sell Herb seeds both retail and wholesale by mail order only. An unusual and interesting listing. Latin and common names given.

**Straw into Gold,** P.O. Box 2904, Oakland, CA 94618. Send 50 cents in stamps for their catalog. They sell supplies, books, and seeds for spinners, weavers, dyers, basket makers, etc.

**Taylors Herb Garden Inc.** 1535 Lone Oak Road, Vista, CA 92083. Telephone (714) 727-3485. Send 25 cents for their catalog. They sell Herb plants both retail and wholesale, at the Gardens and by mail order, shipping by UPS, by truck and by air. They say "Conducted tours every two months over two acres of labeled Herbs, both cultivated and wild." Excellent listing of commonplace and hard to find varieties. Latin and common names given in catalog.

## MICHIGAN

**Fox Hill Farm,** (P.O. Box 7), 434 Michigan Avenue, Parma, MI 49269. Send 50 cents for catalog. Open to visitors Thursdays and Sundays only, noon to 6 p.m. They sell Herb plants at the farm and by mail order; shipping by UPS. Their catalog is exceptionally well presented with many common and hard to find Herbs. Latin and common names given.

## NEW JERSEY

**Mincemoyer Nursery,** W. County Line Road, Jackson, NJ 08527. Telephone (201) 363-3215. Send 25 cents for catalog. They sell Herb plants at the farm and by mail order.

**Thompson and Morgan, Inc.,** 403 Kennedy Blvd., Somerdale, NJ 08083. Telephone (609) 784-8600. Their catalog is free. They sell Herb seeds by mail order both wholesale and retail. This is the U.S. address of this well-known British Seedhouse. Their listing is comprehensive, and Latin and common names are given.

**Well-Sweep Herb Farm,** (Cyrus and Louise Hyde), 317 Mt. Bethel Road, Port Murray, NJ 07865. Telephone (201) 853-5390. Catalog free. Open to visitors Monday to Saturday 9:30 a.m. to 5 p.m. Sunday by appointment only. They sell Herb seeds and Herb plants at the farm and by mail order; shipping by UPS and parcel post. Excellent listing of common and uncommon Herbs.

## OREGON

**Herbs'n Honey Nursery** (Mrs. Chester C. Fisher, Jr.), Route 2, Box 205, Monmouth, OR 97361. Telephone (503) 623-4033. Send 25 cents for catalog. Open year round Monday through Friday, 8 a.m. to 4 p.m. Weekends by appointment. She sells Herb seeds and plants at the farm and by mail order, both wholesale and retail. Good listing of perennial Herb plants, Latin and common names given.

## RHODE ISLAND

**Greene Herb Gardens, Inc.,** Greene, R.I. 02827. Telephone (401) 397-3652. Send stamped self-addressed envelope for seed list which includes both common and hard to find varieties of Herb seeds; they sell by mail order, retail only.

## TEXAS

**Yankee Peddler Herb Farm,** Route 4, Box 76, Hwy. 36 N, Brenham, TX 77833. Telephone (713) 836-4442. Send $1 for a clear and comprehensive catalog, or 25 cents for list. Open to visitors daily 9 a.m. to 5 p.m. except on major holidays. They sell Herb seeds and plants at the farm and by mail order, retail and wholesale. They ship air mail or bus as customers wish. Excellent listing of commonplace and hard to find varieties. Latin and common names given in catalog.

## VERMONT

**Le Jardin du Gourmet** (Raymond Saufroy) West Danville, VT 05873. Telephone (802) 684-2201. Send 25 cents for catalog. He sells Herb seeds and plants by mail order and at the farm after May 15th; shipping by UPS and parcel post. An interesting and unusual catalog which includes some dye plant seeds and three hard to find varieties of shallots.

## CANADA
### Ontario

**Ashby's Garden Centre and Nursery,** R.R. 2, Cameron, Ontario, K0M 1G0. Telephone (705) 359-1115. Send 30 cents for catalog. Open for Herb plant sales Thursday through Saturday. They also sell Herb seeds and plants by mail order. Catalog gives full details. Retail and wholesale sales. Excellent listing of common and many hard to find varieties. Latin and common names given.

**Otto Richter and Sons Ltd.,** P.O. Box 26-G. Goodwood, Ontario L0C 1A0. Telephone (416) 294-1457. Send 50 cents for catalog. Open from 9 a.m. to 5 p.m. seven days a week. Open most days of the year. However, please phone to be certain. They sell Herb seeds by mail order, and Herb seeds and plants at the farm (Canadian customers enquire re shipping plants). Excellent listing, including many unusual and rare Herb seeds. Latin and common names given.

# Appendix "B"

## Books You Are Recommended to Read

*"HOME DYEING WITH NATURAL DYES"* by Margaret S. Furry and Bess M. Viemont
Originally published in 1935 by the United States Department of Agriculture, and for many years unobtainable. Reprinted by Threshold, 443 Sebastopol Avenue, Santa Rosa, California, 95401, U.S.A., in June 1973. To quote from the introduction: *This publication reports the results of tests on about 65 natural dye materials, when used for dyeing cotton and wool cloth. Most of the dyes studied are of vegetable origin . . . only recipes are included for those that produced attractive colors fast to both light and washing.* Price about $1.50

*"DYES FROM PLANTS"* by Seonaid Robertson
Published by Van Nostrand Reinhold, in 1973. This beautifully produced book is a joy to own, for the author not only knows her dyeing, but her English is a delight to read. She is obviously a practical woman . . . so are her explicit directions. Price about $10.

*"YOUR YARN DYEING"* by Elsie G. Davenport
Reprinted by Select Books, 5969 Wilbur Avenue, Tarzana, California, 91356, U.S.A. A must for Handweavers and Spinners who would be Dyers too. Originally written in 1953. The author states in her original preface that, with the exception of one or two natural dyes, nothing is recommended that she has not used herself. She covers natural and commercial dyes, methods for use with wool, silk, cotton, linen and synthetic yarns, and a chapter on color. Price about $4.50

*"THE DYE POT"* by Mary Francis Davidson
Published by the author, Route 1, Gatlinburg, Tennessee, 37738, U.S.A., in 1973, in its seventh printing. Another author who writes from her own experience; her instructions are brief and practical. Price about $2.25

*"THE USE OF VEGETABLE DYES"* by Violetta Thurstan
Published by the Dryad Press, Northgates, Leicester, England, and in the United States at Wood-Ridge, New Jersey, 07075. A "no nonsense" practical little book, particularly good for beginners. Price about $3.50

*"VEGETABLE DYEING"* by Alma Lesch
Published by Watson-Guptill of New York, in 1970. A mine of information, with excellent charts, making vital information easy to locate.

*"NATURAL DYES IN THE UNITED STATES"* by Rita J. Adrosko
She is the Associate Curator, Division of Textiles, Museum of History and Technology. Published by the Smithsonian Institution Press, and obtainable from the Superintendent of Documents, U.S. Government Printing Office, Washington D.C., 20402, U.S.A. Also available from Dover Publications Inc., 180 Varick Street, New York, N.Y., 10014, U.S.A. For anyone interested in the history of dyeing, this book is a must. The second part of the book is a revision of the U.S.D.A. publication "Home Dyeing with Natural Dyes," the first book mentioned on this list. At the end of the book are Appendices, giving much detailed information on colors which can be obtained from various plants, quoted from early nineteenth century authors.

*"SHUTTLE SPINDLE AND DYEPOT"*
A magazine published by the Handweavers Guild of America, on the first day of March, June, September and December. Their business address is 1013 Farmington Avenue, West Hartford, Connecticut, 06107, U.S.A. Membership of the Guild includes subscription to the magazine, and is $12.50 yearly in the U.S.A., $14.50 to foreign countries.

*"SPINNING WITH A DROP SPINDLE"* by Christine Thresh
Published by Thresh Publications (1972), 443 Sebastopol Avenue, Santa Rosa, California, 95401, U.S.A. If you want to try your hand at spinning, directions are clear and simple. Price about $1.25

*"HANDBOOK ON PROPAGATION"*
Published by the Brooklyn Botanic Garden, Volume 13, No. 2 of "Plants and Gardens." All you need to know to propagate plants. This is No. 24 on their list. Price $1.50*

*"DYE PLANTS AND DYEING"*
A handbook published by Brooklyn Botanic Garden, Volume 20, No. 3 of "Plants and Gardens." It has a number of articles on a variety of subjects of interest to the dyer, including one with "35 recipes for the home dyer," which will be found particularly useful; it has no index. This is No. 46 on their list. Price $1.50*

*"NATURAL PLANT DYEING"*
Published by the Brooklyn Botanic Gardens, Volume 29, No. 2 of "Plants and Gardens." Contains a number of articles on a wide variety of subjects of interest to the dyer. Indexed so that plants can be looked up under both their common and Latin names. This is No. 72 on their list. Price $1.50*

*If not available at your local bookstore, Brooklyn Botanic Garden handbooks can be ordered direct from:
Brooklyn Botanic Garden,
1000 Washington Avenue,
Brooklyn, New York, 11225
U.S.A.
Send check payable to Brooklyn Botanic Garden, and order by name and number (post free).